EAT
TO **PREVENT**
AND **CONTROL**
HYPERTENSION

How Superfoods Can Help You Live Disease Free

D1743120

LA FONCEUR

𝓔𝓫
emerald books

$\mathscr{Eb}$

emerald books

Dear reader,

*The aim of **Eat to Prevent and Control Hypertension** is to help reduce your dependence on medicines by providing you with in-depth knowledge of common chronic diseases as well as the best food options that prevent and control diseases naturally.*

Eat healthily, live happily!

Master of Pharmacy, RPh

and Research Scientist

CONTENTS

INTRODUCTION

Nowadays, diabetes, high blood pressure, and arthritis have become quite common. One in every family has one of these diseases. People have started considering these diseases as part of life, which is not good. The lifestyle we are leading today - high intake of processed foods, frequent eating out, smoking, and alcohol, there is a 70% chance that you will have either high blood sugar levels or high blood pressure or both by your 50s.

A disease state in the body means your immune system is constantly busy fighting the disease, soon your immune system loses its effectiveness and becomes weak. If another disease strikes, your immune system is unable to fight, this can have life-threatening consequences. It is very important to start as early as in your 20s to take care of your health. Make your body strong enough to fight any disease naturally.

More diseases mean more medicines. Being from a pharmacy background, I can assure you that dependence on medicines is not good. Medicines prescribed in disease have side effects. To reduce side effects, you are often prescribed with another set of medicines that treat the side effects of your primary medications, but they also have side effects, for which again some other medications are required, so basically, this cycle continues. But there is a solution! You can include foods in your diet that have the same effect as your medications. By regular intake of these foods, you can heal your body and increase your immunity to fight disease naturally.

The objective should be to prevent disease, and preparation starts in your 20s. What you eat in your 20s affects your 50s. To prevent a disease, you must have a thorough knowledge of the disease, such as why it happens? How does it affect your body? What exactly does happen in your body in the event of a disease? What are other health problems that can be caused by a particular disease?

In *Eat to Prevent and Control Hypertension*, you will learn everything about high blood pressure. To prevent this disease, which foods and lifestyle options should you avoid, and which ones should you adopt? What should be your strategy to prevent and control it. What are the foods that mimic your medication's mechanism of action and can help lower your blood pressure? What are the key points that you can follow to prevent and get rid of it?

You will also discover some healthy and tasty recipes that have all the healthy ingredients and still very tasty. These recipes will strengthen your immunity as well as satisfy your taste bud. Get ready for a healthy tomorrow!

UNIT 1

HYPERTENSION: PREVENTION AND CONTROL

1
<u>HYPERTENSION</u>

1.1

EVERYTHING YOU NEED TO KNOW ABOUT HYPERTENSION

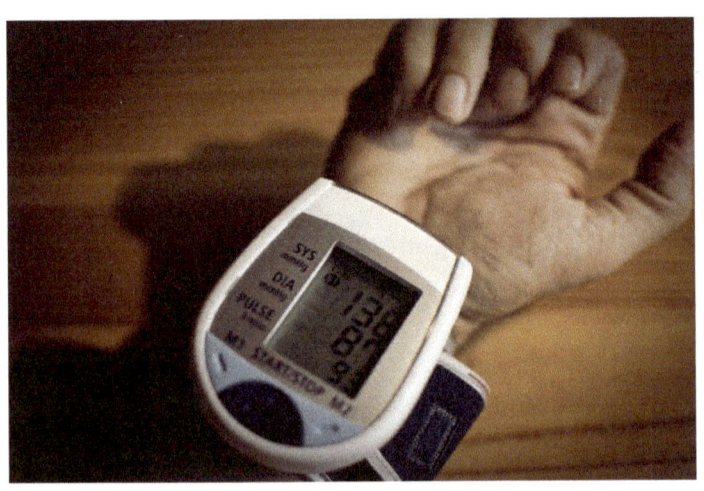

Everything you need to know about hypertension

Hypertension is also known as high blood pressure, is a medical condition that can increase your risk of heart disease, stroke, and other severe health complications. The heart pumps blood into arteries (blood vessels) that carry blood away from the heart to the tissues and organs of the body. Blood pressure is the measure of the force of blood against blood vessel walls. When the force of the blood against the artery walls is too high, it increases the blood pressure, and this condition is known as hypertension.

It is dangerous and needs to be controlled because high blood pressure makes the heart work harder to pump blood out to the organs. This can result in hardening of the arteries (atherosclerosis), stroke, kidney disease, and heart failure.

Blood pressure is expressed by two measurements, maximum and minimum pressures, or a top number and bottom number. Let's see what these numbers are:

Systolic blood pressure - Top number (maximum pressure) tells your systolic pressure. When the heartbeats, it contracts and pumps blood through the arteries to the rest of the body, the contraction creates pressure on blood vessels. This is called systolic blood pressure. Normal systolic pressure is below 120 millimeters of mercury (mm Hg). A reading of 130 mm Hg or higher means high blood pressure.

Diastolic blood pressure - Bottom number (minimum pressure) tells your diastolic pressure. The heart contracts to pump blood to the rest of the body and relaxes before it contracts again. The resting time between beats is when your heart fills with blood and gets oxygen. Diastolic pressure is the pressure of the blood in the arteries when the heart is filling. Normal diastolic pressure is below 80 millimeters of mercury (mm Hg). A reading of 80 mm Hg or higher means high blood pressure.

Among both the numbers, systolic blood pressure is more important than diastolic blood pressure because systolic blood pressure gives the best idea of your risk of having a heart attack or a stroke.

CLASSIFICATION OF BLOOD PRESSURE

Classification of blood pressure for adults aged 18 and older as per ACC/AHA (American College of Cardiology and American Heart Association) and ESC/ESH (European Society of Cardiology and the European Society of Hypertension).

As per ACC/AHA			As per ESC/ESH		
Category	Systolic (mmHg)	Diastolic (mmHg)	Category	Systolic (mmHg)	Diastolic (mmHg)
Normal	Less than 120	Less than 80	Optimal	Less than 120	Less than 80
Elevated	120-129	Less than 80	Normal	120-129	80-84
			High normal	130-139	85-89
Hypertension stage 1	130-139	80-89	Hypertension Grade 1	140-159	90-99
Hypertension stage 2	Equal or more than 140	Equal or more than 90	Hypertension Grade 2	160-179	100-109
			Hypertension Grade 3	Equal or more than 180	Equal or more than 180

As per **ESC/ESH** blood pressure level, **140/90 mm Hg** count as hypertension, whereas new guidelines of **ACC/AHA** count **130/80 mm Hg** as hypertension. It is because of a new study that has found that blood pressure levels between 130/80 mm Hg to 139/89 mm Hg is enough to cause substantial heart and blood vessel complications.

SYMPTOMS OF HIGH BLOOD PRESSURE

In high blood pressure, it is quite possible that you won't experience any symptoms for years or even decades, this is the reason why hypertension count as a silent killer. However, once blood pressure reaches in hypertension stage, a person may have the following symptoms:

- Severe headache (specifically on the back of the head in the morning)

- Chest pain

- Shortness of breath

- Fatigue and confusion

- Vision problems

- Blood in the urine

- Flushing

- Pounding in your chest or neck

The best way to determine whether your blood pressure is high or not is to voluntarily ask your doctor to check your BP when you go for a health check-up. You should get your BP checked every four months. If your blood pressure is elevated, do get checked every month. If you have high blood pressure and you are on medicines, you should check your BP twice a day. The first measurement should be in the morning before eating or taking any medications, and the second in the evening. Each time you measure, take two or three readings at an interval of 1 or 2 mins to make sure your results are accurate.

Let's first understand what actually happens in the body that causes blood pressure to rise.

MECHANISMS THROUGH WHICH BLOOD PRESSURE RISES

1. Abnormalities of the sympathetic nervous system

You must have heard it many times "*do not stress, or else your blood pressure will go up,*" and it's absolutely right." The sympathetic nervous system influences the blood vessels of the body in a dangerous or stressful condition. When you stress out, sympathetic outflow increases. The sympathetic nervous system release the hormones adrenaline and noradrenaline (also known as norepinephrine). These hormones increase the rate of blood pumping from the heart to deliver fresh oxygen to the brain and muscles. The repeated stress means a continuous increase in blood pumping from the heart, which is attained by constricting the blood vessels. The constant constriction of the blood vessels causes the narrowing of the blood vessels and increases the resistance of the blood vessels to blood flow (peripheral resistance). As a result, blood pressure increases.

2. Abnormalities in the intrarenal renin-angiotensin-aldosterone system (RAAS)

This system controls blood pressure by regulating the volume of fluids in the body. When blood flow to the kidneys decreases, the kidneys secrete enzyme renin into blood circulation. Plasma renin then converts angiotensinogen to angiotensin I, which have no direct biological activity. However, an enzyme known as Angiotensin-Converting Enzyme (ACE) converts Angiotensin I to Angiotensin II (the primary hormone responsible for high blood pressure). Angiotensin II is a peptide hormone that causes vasoconstriction, which means it contracts the muscular wall of the vessels, which causes the narrowing of the blood vessels, resulting in increased blood pressure.

RISK FACTORS

Below are the contributory factors that increase your blood pressure:

- Stress

- Obesity

- Excess alcohol

- Smoking

- Family history of high blood pressure

- Excess salt intake

- A diet lack of potassium

- Lack of exercise

- Certain drugs such as NSAIDs, steroids and contraceptive pills

WHY IS HYPERTENSION DANGEROUS?

Hypertension or High Blood Pressure is dangerous because it causes other health complications. Blood vessels are responsible for delivering oxygen and nutrients to vital organs and tissues. Over time, high blood pressure damages the blood vessels. The damaged blood vessels disrupt the blood flow in the body that cause other health problems to arise. The most affected part of the body by hypertension is heart, followed by, brain, kidney, and reproductive system.

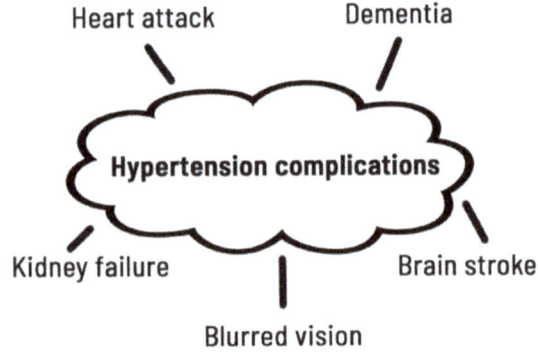

Damage to the arteries

Blood vessels that carry oxygen-rich blood from the heart to the body are called arteries. Arteries are flexible, elastic, and their inner lining is smooth. Blood flows freely and unobstructed through healthy arteries and supply oxygen and nutrients to vital organs and tissues. High blood pressure reduces the elasticity of arteries. It damages their inner lining that makes it easier for dietary fats to collect in the damaged arteries, limiting blood flow throughout your body. These blockages eventually can lead to heart attack and stroke.

Damage to the heart

Hypertension makes your heart pump more frequently, and with more force than a healthy heart, which causes part of your heart (left ventricle) to thicken. An enlarged heart increases your risk of heart attack, heart failure, and sudden cardiac death.

Also, high blood pressure damages the vessels that supply blood to your heart. When blood flow to your heart is obstructed, it can cause arrhythmia (irregular heart rhythms), angina (chest pain), or can cause a heart attack.

Damage to the brain

Our brain depends on nourishing oxygen-rich blood supply to work properly. But high blood pressure can reduce blood and oxygen supply to the brain that can cause several problems, including:

Transient Ischemic Attack (TIA): Hardened arteries or blood clots caused by high blood pressure can temporarily disrupt the blood supply to the brain, which is called a transient ischemic attack (TIA) or mini-stroke. TIA count as a warning of a full-blown stroke.

Stroke: High blood pressure can cause blood clots to form in the arteries, leading to significant blockages in blood flow. Reduced blood flow makes the brain deprived of oxygen and nutrients, causing brain cells to die. This is known as a stroke.

Dementia: Certain types of dementia, such as vascular dementia, is caused by a lack of blood flow in the brain, which may have caused by narrowed/blocked arteries or due to a stroke.

Damage to the kidneys

Kidneys filter excess fluid and waste from the blood. High blood pressure damages the blood vessels in and leading to your kidneys. Damaged vessels obstruct the blood flow to the kidneys and prevent kidneys from effectively filtering waste from your blood, allowing dangerous waste to accumulate. Hypertension is one of the most common causes of kidney failure.

Damage to the eyes

High blood pressure can damage the blood vessels that supply blood to your eyes. Limited blood flow can damage the retina and the optic nerve, leading to bleeding in the eye, blurred vision, and even complete loss of vision.

MECHANISM THROUGH WHICH ANTI-HYPERTENSIVE MEDICINES WORK

The primary purpose of all the anti-hypertensive drugs is to produce vasodilation means to make the blood vessels wider or more open. When blood vessels dilate, blood freely flows through them, causing a fall in blood pressure.

There are different mechanisms through which different classes of drugs achieve vasodilation. Let's see in brief how do these drugs work and how can we produce similar effect through foods, which are safer and have no side effects:

The first-line therapy or most common drugs that are used in hypertension lower the blood pressure by decreasing the salt reabsorption in the kidneys. That means your body now has less salt because more and more salt, along with water, are flushed out from the body through urine. Because you have less fluid in your blood vessels, the pressure inside will be lower. The drugs that work on this mechanism are called Diuretics. Our aim is to include foods that naturally have a diuretic effect. We will later see in detail about the foods that are natural diuretics.

As we discussed above, the Angiotensin II hormone is the main culprit behind high blood pressure. So, the other class of drugs produces vasodilation either by blocking the conversion of angiotensin I to angiotensin II (ACE inhibitors drugs) or by blocking the actions of angiotensin II (Angiotensin receptor blockers drugs). This allows blood vessels to widen and relax, making it easier for blood to flow through, which lowers your blood pressure.

Calcium stimulates the heart to contract more forcefully. A class of drugs called calcium channel blockers limits the rate at which calcium flows into the cells of the heart and blood vessel walls. As a result, blood vessels widen, and your heart doesn't have to work as hard to pump, making it easier for blood to flow through, and your blood pressure lowers.

STRATEGY TO PREVENT AND CONTROL HYPERTENSION

Hypertension is food and lifestyle-related disease that means foods play the most significant part in correcting as well as worsening the condition. Hypertension cannot be solely managed with medicines. Certain diet and lifestyle modifications are must for controlling the disease effectively. Medications prescribed in hypertension have side effects that include erectile impotence, gout, cough, and lack of energy. By adding the right foods for high blood pressure in your diet and avoiding the bad foods, you can effectively lower your blood pressure and can drastically reduce the dose of your blood pressure medicines.

Below are certain ways by which you can lower your blood pressure naturally:

- Eat foods that reduce sodium levels in the body.

- Exclude foods that silently add salt in your body.

- Eat foods that are naturally diuretic.

- Eat foods that reduce fluid retention in the body and increase urine production.

- Eat foods that are rich in magnesium as magnesium is a natural calcium channel blocker.

- Eat foods that are rich in potassium because potassium negates the sodium effect.

- Eat foods rich in nitrates, which convert into nitric oxide in your body. Nitric oxide widens the blood vessels and lowers blood pressure.

- By increasing water intake.

SALT

It must have come to your mind that why is it said to avoid table salt in hypertension? What is the exact relationship between salt and blood pressure? So, let's see why salt is dangerous for blood pressure.

Sodium is the main reason for rising blood pressure, and your table salt is basically a combination of sodium

(40%) and chloride (60%). Are salt and sodium the same? No, not exactly. Sodium is a mineral that occurs naturally in foods. With table salt, you eat sodium in the form of sodium chloride. This is the reason why table salt considered dangerous for high blood pressure.

Other forms of sodium that you consume are

Sodium bicarbonate (baking soda) and

Monosodium glutamate (MSG): Generally used as salt in Chinese foods.

HOW DOES SALT INCREASE BLOOD PRESSURE?

Eating salt raises the amount of sodium in your bloodstream. It affects the performance of your kidneys to remove the water. As a result, your body holds extra water to flush out the extra sodium from your body. This is called fluid retention. The excess fluid in the body puts stress on blood vessels and heart and causes blood pressure to rise.

1.2

10 FOODS THAT RAISE YOUR BLOOD PRESSURE

10 foods that raise your blood pressure

Foods that are high in calories increase cholesterol in your body, which raises blood pressure. Some foods silently add salt to your body and increase your risk of high blood pressure. You don't even release that you are eating salt.

To help you identify these foods, here's a list of 10 foods and drinks that knowingly or silently add salt and cholesterol in your body and can raise your blood pressure.

1. Canned foods or beverages

Canned food products are prepared with lots of salt to preserve the food from decaying, and for taste. This decreases the nutrition of the food and silently add salt in your body. For example, as such, chickpeas are very nutritious and have great health benefits, but if you are using canned chickpeas, then they are harmful to your health. Whenever possible, eat fresh foods instead of canned ones. Even if you are using canned chickpeas or other canned vegetables, wash them properly before use to remove the extra salt. Cans are often lined with the chemical bisphenol A (BPA). Eating foods from cans lined with the chemical bisphenol A (BPA) could raise your blood pressure.

2. Deep-fried foods

Eating deep-fried foods, such as french fries, bagels, and puris, increases your risk of high blood pressure, heart attack, and stroke. Fried foods add a lot of calories and deficient in healthy nutrients. These increase cholesterol levels in the body and raise your blood pressure. You should never reuse the oil that has been used in deep frying, also don't fry at high temperature. When you deep fry, oils break down with each frying, and their composition changes. As a result, a chemical is formed in food that can lead to cancer.

3. Pickles

Eating pickles can raise your blood pressure. Lots of salt and oil are required for preserving pickles for long. Salt and oil stop the food from decaying and keep them edible for long. Adding extra sodium of pickle to your diet causes water retention that puts greater pressure on blood vessels and increases blood pressure. Furthermore, pickles are loaded with excess oil that can increase cholesterol in your body. Cholesterol can narrow the blood vessels that prevent the free-flowing of blood and make it harder for the heart to pump blood through them. As a result, your blood pressure rises.

4. Processed cheese

Processed cheese is high on calories and salt. This means regular consumption can lead to high cholesterol, obesity, and high blood pressure, increasing your risk of heart disease. Avoid eating too much-processed cheese. Instead of processed cheese, go for homemade cottage cheese. You can even eat mozzarella cheese as it contains the lowest salt in comparison to other varieties. Cheese can offer health benefits as they are rich in calcium and vitamins. However, if you have diagnosed with high blood pressure, avoid eating any cheese except homemade low-fat cottage cheese (not the store-bought).

5. Caffeine

Your one cup of coffee may give a boost to start your day in an active mode, but it is not good for blood pressure. Caffeine increases the release of the hormone adrenaline, which is responsible for making you active within some minutes of consumption. However, this same hormone adrenaline causes constriction of blood vessels and increases the rate and force of heart pumping. As a result, pressure inside your blood vessels increases. Don't quit tea or coffee completely, instead gradually reduce the frequency, and soon your body will adapt it. Quitting the caffeine all of a sudden will only make you more crave for it.

6. Dehydration

Keep yourself hydrated. When your body is dehydrated, your brain sends a signal to the pituitary gland to secrete hormone vasopressin, which is an antidiuretic hormone. It constricts the blood vessels. As a result, pressure inside the blood vessels increases, which leads to hypertension. Drink at least eight glasses of water in a day. Drinking enough water in pregnancy is essential, especially if your blood pressure is already high, to prevent toxemia, which is a potentially dangerous pregnancy complication characterized by the onset of high blood pressure.

7. Smoking

Each cigarette you smoke increases your blood pressure. The nicotine in cigarettes is the main reason for high blood pressure. Nicotine stimulates the central nervous system. High blood pressure is caused by both an increase in cardiac output and total peripheral vascular resistance. Nicotine acts as a stimulant in the body. It stimulates the adrenal glands to release more adrenaline. It forces the heart to constrict more forcefully, which affects the heart pumping capacity. It narrows your blood vessels and hardens their walls. As a result, your blood flow gets restricted, which increases peripheral vascular resistance and makes your blood more likely to clot. If you smoke, quit it as soon as possible—people who quit smoking live longer than people who smoke.

8. Alcohol

Any alcoholic beverage can increase your blood pressure. Alcohol interferes with the effectiveness of your anti-hypertensive drugs. In fact, even one drink can change the way your blood pressure medications work and increase the side effects of blood pressure medications. Alcohol increases blood pressure by increasing cortisol (stress hormone) levels. Alcohol stimulates the release of vasoconstrictor angiotensin II and induces inflammation that inhibits nitric oxide production in the body. Alcohol is high in calories and may contribute to unwanted weight gain, which is a risk factor for high blood pressure. If you want to prevent high blood pressure, quit drinking completely.

9. Sugar

It might surprise you, but high sugar intake is linked with hypertension. Not only because it leads to obesity, but it causes high blood pressure through a different mechanism too. Sugar elevates uric acid in the body that induces oxidative stress and decreases endothelial nitric oxide availability. It also activates both renin activity and angiotensin activity in the kidney. Nitric oxide suppression, as well as enhancement of angiotensin activity, constricts the blood vessels, leading to high blood pressure. Sugar also gives empty calories and increases cholesterol in the body. Increased cholesterol causes plaque to build up inside your arteries, and arteries become hardened and narrowed. As a result, your heart has to work much harder to pump blood through them and cause an increase in your blood pressure.

10. Packaged snacks

Potato chips, popcorns, sev, banana chips, and other packaged snacks have lots of salt in it. More the flavors, the more the salt content. Flavors like salted, cream, and onion have more salt content. Additionally, these are high in saturated and trans fats, sugar, and other low-fiber carbohydrates. Consuming too much of packaged snacks increases bad cholesterol (LDL) in the body. High LDL levels may lead to the formation of cholesterol plaque in the blood vessel that constricts the free-flowing of blood and may eventually lead to the development of coronary heart disease. You don't need to stop eating snacks completely, but do limit your servings and frequency. Look for plain version instead of flavored.

1.3

10 FOODS THAT LOWER BLOOD PRESSURE JUST LIKE ANTIHYPERTENSIVE DRUGS

10 foods that lower blood pressure just like antihypertensive drugs

Certain foods naturally lower blood pressure. Adding these foods in your diet can significantly reduce your risk of high blood pressure. A diet low in sodium and rich in potassium, magnesium, nitrate, and fiber can help prevent and control blood pressure.

Below are the top 10 foods that work just like antihypertensive drugs and lower your blood pressure:

1. Leafy green vegetables

Leafy green vegetables such as spinach, kale, fennel, and cabbage are very dense in nutrition. These are packed with potassium, magnesium, nitrates, and full of fiber. Low potassium intake is one of the risk factors in developing high blood pressure. The potassium content of leafy green vegetables flushes out sodium. The rich magnesium content act as calcium channel antagonist, modulating vascular tone and reactivity that dilates blood vessels. As a result, blood flows without any restriction through your blood vessels, and your blood pressure decreases.

Moreover, nitrate of leafy greens induces vasodilation. Nitrate converts into nitrite in your mouth by oral commensal bacteria. It then converts in nitric oxide in your blood, which relaxes the smooth muscle of your blood vessels and dilates them that allows free-flowing of blood. Prevent using antibacterial mouthwash as it inhibits nitrate to nitrite conversion, and you don't get the blood pressure-lowering benefit of nitric oxide.

2. Beetroot

Beetroot is a well-known potent vasodilator. It contains high levels of dietary nitrate; the body converts the nitrate in this vegetable into nitric oxide. Nitric oxide relaxes and dilates blood vessels, so lowering blood pressure. For the full benefit, drink a glass of raw beetroot juice because raw beetroot has more potency than cooked one. Research indicates that within hours of drinking raw beetroot juice, it lowers systolic blood pressure.

3. Garlic

Eating garlic every day may help reduce your blood pressure. The organosulfur compounds of garlic promote vasodilation and lower blood pressure. Garlic lowers blood pressure through various mechanisms of action. Dietary intake of garlic boosts hydrogen sulfide production and regulation of endothelial nitric oxide, which induces smooth muscle cell relaxation and vasodilation; as a result, blood pressure drops. Garlic also blocks angiotensin-II production by inhibition of the angiotensin-converting-enzyme (ACE).

Allicin is an organosulfur compound that releases when garlic is crushed or chopped. Allicin is highly unstable. Cooking speeds up the degradation of allicin, and microwaving destroys it completely.

Eat one clove of freshly crushed garlic empty stomach every day, and it will lower your blood pressure naturally. But keep in mind that raw garlic is quite pungent and can cause burn, so don't hold it in your mouth for long periods. Too much garlic may cause irritation and digestive upset. If you see the irritation, limit the frequency to 2 to 3 times a week.

4. Cucumber

Cucumber is rich in potassium, which plays an important role in regulating blood pressure. The excess sodium in your body reduces the ability of your kidneys to remove the water. This makes your body hold fluids that raises your blood pressure. Potassium lowers blood pressure by balancing out the adverse effects of sodium. The more potassium you eat, the more sodium you lose through urine. Also, cucumber is a diuretic. It flushes out sodium from the body by increasing your urine production and maintain fluid balance in the body that helps keep blood pressure in check. Use cucumber in salad, raita, or have cucumber juice.

5. Banana

What can be a better source of potassium than the well-known and probably most blood pressure friendly fruit banana! Eating potassium-rich foods decrease sodium levels in your body. It reduces fluid retention and blood pressure by increasing urine production. Potassium also acts as a vasodilator that helps ease tension in your blood vessel walls, which helps lower blood pressure. Banana is one of the healthiest fruits because it is very low in calories and has higher water content.

6. Lemon water

Lemon is an excellent remedy for hypertension as it helps keep blood vessels soft and pliable, making them flexible by removing any rigidity. This keeps the blood pressure low. Lemon is rich in vitamin C that works as an antioxidant and has a diuretic effect. It removes excess fluid from your body, which lowers the blood pressure. Furthermore, vitamin C helps protect the levels of nitric oxide in the body that relaxes blood vessels and contributes to maintaining normal and healthy blood pressure. Taking a glass of warm lemon water every morning on an empty stomach helps you keep the high blood pressure at bay. If you are on blood pressure medicines, do consult your doctor and pharmacist before including citrus fruits like lemon in your diet as citrus fruits can interact with your medications, especially calcium antagonist drugs.

7. Honey

Honey contains antioxidant compounds that are linked to lower blood pressure. Obesity and unhealthy lifestyle cause oxidative stress in your body that reduces vasodilatory agent nitric oxide available in the body. The antioxidants present in honey help to keep the nitric oxide levels high in the body by reducing oxidative stress in the body. Nitric oxide relaxes your blood vessels causing vasodilation, which helps to lower the blood pressure. Take one tablespoon of honey daily or add it in your morning lemon water. Make sure you eat organic honey, not the processed ones. Eat it raw, don't heat the honey. Heating honey destroys the beneficial enzymes, vitamins, and minerals of the honey.

8. Nuts

Nuts such as almonds, cashew nuts, and walnuts are rich in magnesium, fiber, and protein. Magnesium is an electrolyte that helps lower high blood pressure. Magnesium is a natural calcium channel blocker, it stimulates the production of vasodilators nitric oxide and prostacyclins. These vasodilators relax the blood vessels and lower blood pressure. Nuts also contain heart-healthy fats that lower cholesterol levels. Remember to get magnesium from your food, not from supplements, to avoid any risk of overdose. Make sure to eat nuts every day.

9. Fenugreek seeds

Drinking fenugreek water can be one of the most effective ways for you to maintain healthy blood pressure. Fenugreek leaves and seeds contain a high amount of dietary fiber. A diet rich in fiber has been linked to steady levels of blood pressure. Dietary fiber is tough to digest. It forms a viscous gel in the intestine that makes it harder for sugars and fats to absorb in the bloodstream and decreases cholesterol levels in the body and prevent weight gain. Furthermore, fenugreek leaves and seeds contain low levels of sodium, which makes it an ideal food for people with high blood pressure.

Take two teaspoons of fenugreek seeds and soak them in a glass of water overnight. The next morning remove the seeds from water, drink the fenugreek water on an empty stomach. Crush the seeds to a fine paste and use it in cooking. Do this for at least two to three months and see the positive result yourself.

10. legumes

Legumes such as lentils, chickpeas, kidney beans, soybeans, and others, are rich in potassium, magnesium, and fiber. These nutrients maintain healthy and normal blood pressure. Potassium and magnesium in legumes prevent fluid retention, decrease sodium levels by increasing urine production. The soluble fiber in the legumes gets attached to cholesterol particles and takes them out of the body that helps to reduce overall cholesterol levels. This reduces the risk of weight gain and improves vascular health.

IMPLEMENTATION

Now that you know everything about hypertension, how blood pressure works, what exactly goes wrong in your body in hypertension condition, what are the risk factors, what you should avoid and what you should start adding in your diet? Now the question arises, how to implement it?

If you are a healthy person and have no hypertension condition and no family history of hypertension, start with adding foods in your diet that can help you prevent hypertension in the future. Start with adding above mentioned foods that are most of your likening, gradually add those foods in your diet that you do not like much but make sure to add them, gradually you will be habitual. You must add sweet potato, onion, and pomegranate too in your diet; these foods are also very effective in preventing high blood pressure. Limit your consumption of food that increases your risk of high blood pressure. If you smoke, quit it. Limit your alcohol consumption.

If you are a healthy person but have a family history of hypertension, strictly avoid eating known and silent salt-rich foods. With a family history of hypertension, you are susceptible to high blood pressure. Quit smoking and drinking any alcoholic beverages completely, don't increase your risk of high blood pressure. Eat foods as mentioned above that naturally prevent hypertension, also include sweet potato, pomegranate in your diet, but if you have a family history of diabetes too, then limit your consumption of sugar as well as sweet fruits.

If you are a person with high blood pressure, you must eat potassium-rich foods. The best way is to note down all the above mentioned preventive foods, ask your doctor and pharmacist:

I want to add these foods in my diet, is it safe to add all of them? Are there any foods that can interact with my medications? They will provide you the best advice considering your blood pressure levels and other health complications. Let them know if you have other health problems like diabetes and arthritis. If you have diabetes, then limit the consumption of high glycemic foods. If you have arthritis, limit your citrus fruit consumption. After including these foods for three months, ask your doctor, has my blood pressure improved? Do I need the same dose of medicines or lower dose will work? Keep in mind once you have controlled your blood pressure, continue to consume foods that prevent high blood pressure. It should be a lifetime habit, not short-term therapy.

KEY POINTS

✓ Consume less salt. It promotes fluid retention that increases blood pressure.

✓ Avoid foods that silently add salt in your body.

✓ Eat foods low in fat and calories.

✓ Do yoga.

✓ Be active. Physical activity reduces water retention by making you sweat and increasing blood flow to the tissues.

✓ Drink plenty of water, especially on the day you consume more salt. Dehydration can increase your risk of water retention.

✓ Don't stress out.

✓ Eat whole grains instead of refined ones.

✓ Add barley flour to your whole wheat flour in a ratio of 2:10. Add 200 gm of barley flour to 1 kg of whole wheat flour.

UNIT 2

DIET PLAN

Diet Plan for

Hypertension

Diabetes + Hypertension

Hypertension + Arthritis

DIET PLAN TO CONTROL HYPERTENSION

Diet plan to control hypertension

- Add one tablespoon of honey in warm lemon water and drink it on an empty stomach.

- Eat one crushed garlic on an empty stomach every day (if you experience mouth ulcer or heat in the body, reduce the frequency to 5 times a week).

- Soak one tablespoon of fenugreek seeds overnight in 250 ml of water. The next morning chew these seeds and drink the fenugreek water. Do this thrice a week.

- Eat one banana every day, especially if you are taking hypertension medicines.

- Drink green tea and add lemon to it.

- In season, eat multigrain beetroot paratha (see the recipes section). Have 50 to 100 ml of beetroot juice every day.

- Eat a handful of overnight soaked nuts.

- Add barley flour to your whole wheat flour in a ratio of 2:10. Add 200 gm of barley flour to 1 kg of whole wheat flour. Beta-glucan of barley is very effective in preventing hypertension and prevent weight gain.

- In winter, eat plenty of spinach, kale, and chenopodium (bathua).

- Eat one flax seeds laddoo (see in the recipes section) every day.

- Eat plenty of sweet potatoes, especially the purple-fleshed sweet potatoes.

- Eat plenty of legumes such as lentils, chickpeas, kidney beans, and soybeans for potassium, magnesium, and fiber.

- Drink plenty of water, especially on the day you consume more salt.

- Eat fresh fruits, like apples, oranges, and bananas.

- Drink low-fat cow's milk boiled with turmeric powder every night.

- Eat mixed veg raita (see in the recipes section) with lunch.

DIET PLAN TO CONTROL DIABETES + HYPERTENSION

Diet plan to control diabetes + hypertension

- Drink warm lemon water on an empty stomach.

- After half an hour, eat soaked fenugreek seeds and drink the fenugreek water. Do this every day.

- After one hour, eat one crushed garlic. Do this every day.

- Have green tea with added lemon and basil leaves.

- Eat a handful of overnight soaked nuts.

- Drink about 2-3 liters of water in a day.

- Add barley flour to your whole wheat flour in a ratio of 2:10. Add 200 gm of barley flour to 1 kg of whole wheat flour.

- Eat one banana, especially if you are taking hypertension medicines. Don't remove the banana from your diet just because you have diabetes. Cut the sugar intake in tea and other high glycemic fruits.

- In season, drink 50 ml to 100 ml of fresh bitter gourd juice every day.

- In season, eat multigrain beetroot paratha (see the recipes section). Have 50 to 100 ml of beetroot juice every day.

- Eat sprouts every day.

- Add flax seeds in dough or add them in a yogurt fruit salad.

- Eat bottle gourd, carrot, and ridge gourd.

- Replace potatoes with sweet potatoes and eat them in moderation.

- Drink low-fat cow's milk boiled with turmeric powder at night.

- Eat plenty of spinach, kale, cabbage, and chenopodium (bathua).

- Increase legumes consumption, including lentils, chickpeas, kidney beans, and soybeans.

- Eat vitamin C foods such as amla, lemon, orange, and capsicum.

- Use olive oil, canola oil, and mustard oil in cooking.

DIET PLAN TO CONTROL HYPERTENSION + ARTHRITIS

Diet plan to control hypertension + arthritis

- Eat one crushed garlic on an empty stomach every day (if you experience mouth ulcer or heat in the body, reduce the frequency to 5 times a week).

- After half an hour, eat soaked fenugreek seeds and drink the fenugreek water. Do this thrice a week.

- Drink green tea mixed with freshly crushed ginger.

- Eat a handful of overnight soaked dry fruits, including two walnuts, two dried figs, five almonds, four cashew nuts, four pistachios, and four raisins. Eat them daily.

- Drink low-fat turmeric milk at night.

- Eat one banana every day, especially if you are taking hypertension medicines.

- Add barley flour and soybean flour to your whole wheat flour in a ratio 2:1:10. Add 200 gm of barley flour and 100 gm of soybean flour in 1 kg of whole wheat flour.

- Eat plenty of horse gram (kulthi). Make horse gram dal or powder it and use with your whole wheat flour or add the powder in buttermilk.

- In season, eat multigrain beetroot paratha (see the recipes section). Have 50 to 100 ml of beetroot juice every day.

- In winter, eat plenty of fresh turmeric roots, spinach, kale, fenugreek leaves, chenopodium, and sweet potatoes, especially the purple-fleshed sweet potatoes.

- Eat one flax seeds laddoo (see in the recipes section) every day.

- Eat mixed veg raita (see in the recipes section) with lunch.

- Eat plenty of legumes such as lentils, chickpeas, kidney beans, and soybeans for potassium, magnesium, and fiber.

- Drink plenty of water, especially on the day you consume more salt.

- Use extra virgin olive oil (only for shallow fry), canola oil or mustard oil in cooking, avoid sunflower oil and corn oil, which are rich in omega-6 fats.

- Eat fresh fruits, like apples, oranges, and bananas.

UNIT 3

RECIPES

Healthy and tasty recipes to boost your health

BREAKFAST

MULTIGRAIN BEETROOT PARATHA

MIXED VEG RAITA

Breakfast

MULTIGRAIN BEETROOT PARATHA

Multigrain beetroot paratha

<u>To make 8 parathas</u>

Ingredients:

Grated beetroot: 1 cup

Grated bottle gourd: 1 cup

Whole wheat flour: 1 cup

Oats flour: 1 cup

Gram flour/chickpea flour: ½ cup

Barley flour: ¼ cup

Amaranth flour: ¼ cup

Salt: To taste

Ginger garlic paste: 1 tablespoon

Jaggery: 2 tablespoons

Coriander powder: 1 teaspoon

Red chili powder: 1 teaspoon

Garam masala: 1 teaspoon

Turmeric: 1 teaspoon

Asafoetida: a pinch

Sesame seeds: 1 tablespoon

Crushed fenugreek seeds: 1 teaspoon

Olive oil: 3 tablespoons

Yogurt: 2 tablespoons (to knead)

Method:

1. Mix all ingredients along with one tablespoon of olive oil in a large mixing bowl.

2. Knead with yogurt and make a stiff dough.

3. Divide the dough into 8 equal balls.

4. Take one piece of the dough ball and dip it in the dry whole wheat flour, dust off the excess flour.

5. Use a rolling pin to roll the dough into a circle.

6. Heat the pan/griddle (tawa) on medium-high flame.

7. Place the paratha on the griddle. Cook for about a minute or cook until the paratha begins puffing up from the base at some places.

8. Flip the paratha and spread 3-4 drops of olive oil. Cook for 2 minutes until it turns light brown.

9. Flip the paratha again and top with 3-4 drops of olive oil, spreading it evenly over the surface. Gently press the paratha with the flat spatula to help the paratha cook evenly.

10. Once you begin to see brown spots on both sides of the paratha, transfer the paratha to a serving plate. Your paratha is ready. Similarly, make all the parathas.

11. Enjoy multigrain beetroot paratha with mixed veg raita.

MIXED VEG RAITA

Mixed veg raita

Ingredients:

Yogurt: 200 grams

Grated beetroot: 1 tablespoon

Finely chopped onion: ¼ cup

Finely chopped tomato: ¼ cup

Finely chopped cabbage: ¼ cup

Finely chopped cucumber: ¼ cup

Brown sugar: 2 tablespoons

Black Salt: 1 teaspoon

Black pepper powder: ½ teaspoon

Red chili powder: ½ teaspoon

Cumin powder: 1 tablespoon

Method:

1. Blend the yogurt and make it smooth.

2. Mix all the veggies except beetroot in yogurt.

3. Add black salt, brown sugar, black pepper powder, red chili powder, and cumin powder in yogurt. Mix well.

4. Keep the raita in the fridge for half an hour.

5. Garnish with grated beetroot.

6. Eat with multigrain beetroot paratha.

NOTE FROM LA FONCEUR

Dear Reader,

Thank you for reading *Eat to Prevent and Control Hypertension*. I hope you have found this book helpful.

If you liked the book, please leave a review online. Help other health-conscious readers find this book by telling them why you enjoyed reading. Your help in spreading awareness will be gratefully received.

Join my mailing list at www.eatsowhat.com/mailing-list

If you are looking for a permanent solution to your hair problems, read *Secret of Healthy Hair*.

Learn how a vegetarian diet can be the solution to a disease-free healthy life in Eat So What! series- *Eat So What! The Power of Vegetarianism* and *Eat So What! Smart Ways to Stay Healthy*.

All of my books are available in eBook, paperback, and hardcover editions.

Regards

La Fonceur

REFERENCES

1. Soheil Z, Habsah A, "A Review on Antibacterial, Antiviral, and Antifungal Activity of Curcumin." Biomed Res Int. 2014; 2014: 186864.
2. Silagy C, Neil A, "Garlic as a lipid lowering agent--a meta-analysis." R Coll Physicians Lond. Jan-Feb 1994;28(1):39-45.
3. Matthias B, Mandy S, "Fiber and magnesium intake and incidence of type 2 diabetes: a prospective study and meta-analysis." Arch Intern Med. 2007 May 14;167(9):956-65.
4. Karin R, Toben C, Fakler P, "Effect of garlic on serum lipids: an updated meta-analysis." Nutr Rev. 2013 May;71(5):282-99.
5. Holly L, Sharon A, "Garlic and onions: Their cancer prevention properties." Cancer Prev Res (Phila). 2015 Mar; 8(3): 181–189.
6. Ranade M, Mudgalkar N, "A simple dietary addition of fenugreek seed leads to the reduction in blood glucose levels: A parallel-group, randomized single-blind trial." Ayu. 2017 Jan-Jun; 38(1-2): 24-27.
7. Calado A, Neves M, "The Effect of Flaxseed in Breast Cancer: A Literature Review." Front Nutr. 2018; 5: 4.
8. Chikako M, Taeko K, "Effects of glycolipids from spinach on mammalian DNA polymerases." Biochem Pharmacol. 2003 Jan 15;65(2):259-67.
9. Mondal S, Varma S, "Double-blinded randomized controlled trial for immunomodulatory effects of Tulsi (Ocimum sanctum Linn.) leaf extract on healthy volunteers." Ethnopharmacol. 2011 Jul 14;136(3):452-
10. Malouf M, Grimley E, "Folic acid with or without vitamin B12 for cognition and dementia." Cochrane Database Syst Rev, 2003;(4): CD004514.
11. Radzeviciene L, Ostrauskas R, "Adding Salt to Meals as a Risk Factor of Type 2 Diabetes Mellitus: A Case–Control Study." Nutrients. 2017 Jan; 9(1): 67.
12. Li J, Zhang N, "Non-steroidal anti-inflammatory drugs increase insulin release from beta cells by inhibiting ATP-sensitive potassium channels." Br J Pharmacol. 2007 Jun; 151(4): 483–493.
13. Kato Y, Domoto T, "Effect on Blood Pressure of Daily Lemon Ingestion and Walking." J Nutr Metab. 2014; 2014: 912684.
14. Dehkordi F, Kamkhah A, "Antihypertensive effect of Nigella sativa seed extract in patients with mild hypertension." Fundam Clin Pharmacol, 2008 Aug;22(4):447-52.
15. Diego A, Ocampo B, "Dietary Nitrate from Beetroot Juice for Hypertension: A Systematic Review." Biomolecules. 2018 Dec; 8(4): 134.
16. Ried K, Fakler P, "Potential of garlic (Allium sativum) in lowering high blood pressure: mechanisms of action and clinical relevance." Integr Blood Press Control. 2014; 7: 71–82.

17. Aluko E, Olubobokun T, "Honey's Ability to Reduce Blood Pressure and Heart Rate in Healthy Male Subjects." Frontiers in Science, 2014

18. Jovanovski E, Bosco L, "Effect of Spinach, a High Dietary Nitrate Source, on Arterial Stiffness and Related Hemodynamic Measures: A Randomized, Controlled Trial in Healthy Adults." Clin Nutr Res. 2015 Jul; 4(3): 160-167.

IMPORTANT TERMINOLOGY

Arteries: Arteries are blood vessels that carry oxygen-rich blood from the heart to the body.

Vascular: Vascular is related to vessels that carry blood in the body.

Vasoconstriction: Vaso means vessels, so vasoconstriction means narrowing of the blood vessels.

Vasodilation: Dilation or widening of blood vessels.

Bioavailability: The actual proportion of a substance that reaches the blood circulation after it introduced into the body to show its effect.

Antioxidants: Body has antioxidants, which neutralize free radicals by inhibiting the oxidative process that forms free radicals.

Oxidative stress: When free radicals outnumber the naturally occurring antioxidants, it results in oxidative stress. This imbalance leads to cell and tissue damage, including DNA, protein, and lipids. Damage to your DNA increases your risk of chronic diseases such as cancer, rheumatoid arthritis, diabetics, stroke, and aging.

ABBREVIATIONS

BP- Blood pressure

LDL- Low-density lipoproteins

HDL- High-density lipoprotein

COX- Cyclooxygenase

PG- Prostaglandin

RAAS- Renin–angiotensin–aldosterone system

ABOUT THE AUTHOR

La Fonceur is the author of the book series *Eat So What!* and *Secret of Healthy* Hair, a dance artist, and a health blogger. She has a master's degree in Pharmacy. She specialized in Pharmaceutical Technology and worked as a research scientist in the research and development department. She has published an article titled 'Techniques for Producing Biotechnology-Derived Products of Pharmaceutical Use' in Pharmtechmedica Journal. She is also a registered pharmacist. Being a research scientist, she has worked closely with drugs. Based on her experience, she believes that one can prevent most of the diseases with nutritious vegetarian foods and a healthy lifestyle.

ALL BOOKS BY LA FONCEUR

Full-length books:

Mini editions:

Hindi editions:

CONNECT WITH LA FONCEUR

Instagram: @la_fonceur | @eatsowhat

Facebook: LaFonceur | eatsowhat

Twitter: @la_fonceur

Amazon Author Page:

www.amazon.com/La-Fonceur/e/B07PM8SBSG/

Bookbub Author Page:
www.bookbub.com/authors/la-fonceur

Sign up to my website to get exclusive offers on my books:

Blog: www.eatsowhat.com

Website: www.lafonceur.com/sign-up

Lightning Source UK Ltd.
Milton Keynes UK
UKHW050816150221
378796UK00007B/99

9 781034 419853